MEASURING TIME

by Meg Gaertner

Cody Koala

An Imprint of Pop!

popbooksonline.com

abdobooks.com
Published by Pop!, a division of ABDO, PO Box 398166, Minneapolis, Minnesota 55439.

Printed in the United States of America, North Mankato, Minnesota

102019
012020

THIS BOOK CONTAINS RECYCLED MATERIALS

Cover Photo: Shutterstock Images
Interior Photos: Shutterstock Images, 1, 5 (top); iStockphoto, 5 (bottom left), 5 (bottom right), 7, 9, 11, 13, 15, 17, 19, 21

Editor: Meg Gaertner
Series Designer: Jake Slavik

Library of Congress Control Number: 2019942413

Publisher's Cataloging-in-Publication Data
Names: Gaertner, Meg, author.
Title: Measuring time / by Meg Gaertner
Description: Minneapolis, Minnesota : Pop!, 2020 | Series: Let's measure | Includes online resources and index.
Identifiers: ISBN 9781532165566 (lib. bdg.) | ISBN 9781532166884 (ebook)
Subjects: LCSH: Time--Juvenile literature. | Time--Perception--Juvenile literature. | Measurement--Juvenile literature. | Mathematics--Juvenile literature. | Clocks and watches--Juvenile literature.
Classification: DDC 530.813--dc23

Hello! My name is

Cody Koala

Pop open this book and you'll find QR codes like this one, loaded with information, so you can learn even more!

Scan this code* and others like it while you read, or visit the website below to make this book pop.

popbooksonline.com/measuring-time

*Scanning QR codes requires a web-enabled smart device with a QR code reader app and a camera.

Table of Contents

Chapter 1

What Is Time?

Time is the nonstop series of events. Past events happened before. Present events are happening now. Future events have not happened yet.

past

Watch a video here!

present

future

Chapter 2

Units of Measurement

People measure time to know when things happen. They measure time in **hours** and **minutes**. People can tell the time by looking at a clock.

Learn more here!

Analog clocks show 12 numbers in a circle. These clocks have hands that move and point. The short hand points to the hour. The long hand points to the minute.

short hand
long hand
QUARTZ

Digital clocks show numbers separated by a colon. The numbers to the left of the colon tell the hour. The numbers on the right tell the minute.

Some digital clocks say if it is a.m. or p.m. Times that are a.m. are in the morning. Times that are p.m. are in the afternoon or evening.

colon
PM
hour
minute

Chapter 3

How to Tell Time

People read an **analog clock** by looking at its hands. First, look where the short hand is pointing. It tells the **hour**.

Learn more here!

Then, look where the long hand is pointing. It tells the **minute**. If the long hand points up at the 12, it is right on the hour. A new hour is beginning.

5:00

If the long hand points down at the 6, it is halfway through the hour. Thirty minutes have passed. There are still 30 minutes to go until the next hour.

8:30
long hand

The long hand can point at the other numbers too. Each number shows a five-minute **interval**. It tells how many minutes have passed in the hour.

In the past, people told time by looking at shadows made by the sun.

6:00

6:05

6:10

6:15

6:20

6:25

Chapter 4

Measure It!

Here is an **analog clock**. The short hand is pointing toward the 3. The long hand is pointing at the 7. What time is it?

Complete an activity here!

Making Connections

Text-to-Self

When have you needed to know the time? How did you learn the time?

Text-to-Text

Have you read other books about measuring time? What did you learn?

Text-to-World

Mitsuko looks at an analog clock. The short hand points to the 2. The long hand points to the 6. What time is it?

Glossary

analog clock – a clock that gives the time by the positions of moving hands.

digital clock – a clock that gives the time in numbers.

hour – a period of time that lasts 60 minutes. There are 24 hours in a day.

interval – a period of time.

minute – a period of time that lasts 60 seconds. There are 60 minutes in an hour.

Index

Online Resources

popbooksonline.com

Thanks for reading this Cody Koala book!

Scan this code* and others like it in this book, or visit the website below to make this book pop!

popbooksonline.com/measuring-time

*Scanning QR codes requires a web-enabled smart device with a QR code reader app and a camera.